KEYS TO THE CLUBHOUSE

KEYS TO THE CLUBHOUSE

7 HACKS FOR A LESS STRESSFUL MOVE
INTO THE **CLUBHOUSE APP**

TASHA HART

KEEN VISION PUBLISHING

Cover Design: by Triv ilovedesignsbytriv@gmail.com

Printed in the United States of America
Keen Vision Publishing, LLC
www.publishwithKVP.com
ISBN: 978-1-948270-99-1

For every creative looking for a virtual space to call home.

TABLE OF CONTENTS

Technology and social media have brought power back to the people.

MARK MCKINNON

INTRODUCTION

The ability to set up a new social media presence and manage it effectively is a challenge no matter which virtual neighborhood you embark on. This was true for me and explained my many moves from one platform to the next, trying to find my forever virtual home. I once thought I couldn't settle down in a virtual home due to some fault of my own. I blamed my lackluster attempts to grasp all the technical stuff on my inability to figure out this thing they called an algorithm. Then, I questioned my hospitality skills once I reviewed the number of visitors who came by my social media space. For years, I gave up on ever finding a virtual space I could truly call home. I settled for vacation spots and rentals, thinking I would never wrap my mind around this new virtual world enough to take ownership of it for myself. This was my situation until I was introduced to a new application called Clubhouse.

I wrote this book for everyone who has experienced the stress of moving into a new social media platform. The countless hours unpacking the features and understanding the new space can be difficult. This short read is my way of providing you with the movers, un-packers, and decorators so you can relax. Welcome home, neighbor!

Hack One

THE REALTOR

Surprisingly, I did not know I was in the market for a new social media home. It was not something I intentionally embarked upon a journey to discover. I had made up my mind that I would never settle down. I would simply enjoy the various locations around the virtual world and experience life on the go with family and friends. For me, it felt like I was always a visitor but never the host. Whenever I moved into a new virtual neighborhood, I would find that my neighbors moved into the virtual space years before me. My new neighbors would welcome me over from time to time, but for the most part, everyone stayed in their comfortable circles. It was challenging to break through those circles and become a part of a scene without years of work each and every time. This did not make the virtual space a bad one. However, it was not exactly welcoming.

Over the years, I began to desire a deeper connection to my virtual community. I wanted something beyond surface-level, where my latest post wasn't measured by how many followers viewed it or gave it a thumbs-up. I discovered that each virtual platform was more focused on features and getting as many billboards to display ads in our virtual neighborhood as possible. Some platforms even put billboards directly in your front yard. Somehow, my neighbors and I carried on with life on these platforms and navigated around the noise.

Before I knew that is what he was to me, my realtor came to me one day, recognizing my wandering ways, and asked me a question. "Have you heard of this app called Clubhouse?" He asked. To tell you the truth, I had not heard of this new app. Honestly, I was not interested in another social media platform that would connect me with more apologies than actual benefits and more features than friends. I had enough apps on my phone to keep me swiping right for days. I did not see the need for one more app filled with promises they could not deliver. People were finally starting to come around and steal my newspaper from the driveway at my current residence. I was willing to take the interaction where I could get it at that point. My virtual Realtor, however, was not discouraged by my unexciting answer. Justin Cappon with Justin Cappon Pro Services insisted I let him show me this new virtual property. Justin and I walked through the front door of that virtual space, and my life changed forever. Justin had found a home with this new app and couldn't help but share his enthusiasm with everyone he knew. I am grateful that he decided to extend that invitation to me. Now, I feel obligated to do the same.

Justin Cappon did his research. Knowing my desire to be in a community of heart-centered individuals with the potential to grow personally and professionally, he located what would become my future virtual home. So here's hack number one: Find a great virtual realtor. A realtor is simply someone who helps you to acquire buildings and land. In this case, we are talking about virtual real estate. Like Justin Cappon, your virtual realtor should be very aware of your lifestyle needs and desires. Also, your realtor should be well-informed about the space you may potentially be moving into in order to determine if that space is a good fit for you. The realtor will help you determine how much "home" you can afford so you can find a property that fits your lifestyle.

Knowing how much virtual "home" you can afford comes down to understanding what it will cost you to move from your current "home" to another. You may not be able to justify the cost at this time, depending on what your existing home means to you. For me, I had been between "homes," if you will, and that afforded me the flexibility to move while discovering how my new "home" would affect my life moving forward. Would this be a permanent residence or just another space I visited from time to time, depending on my agenda and needs for the day, month, or year? Not everyone will find a need or desire to move, but I encourage you to discuss the options with your newfound realtor. Sometimes, just being presented with the options can surprise you. You may not need a mansion with multiple rooms and a four-car garage but instead a studio apartment. That could be the place that puts you in the neighborhood of like-minded individuals who support your dream to become and do what you have always desired. So, what are you waiting for? Go ahead and at least explore the options.

Again, at the time, I didn't know that I was in the market for a new virtual "home," but when my realtor invited me to look at property in this new virtual community called Clubhouse, I quickly fell in love. You may be thinking the same thing. You are comfortable in your current space. You know where everything is, you have gotten used to the way things are, and you have accepted that things may never change. Everyone won't take the leap and move. However, it never hurts to at least take a look. When presented with the options, based on my current needs and desire to be in a community that supports me, I took a chance. If someone has mentioned the Clubhouse app to you and you're thinking about exploring and possibly moving into this community, keep reading and consider becoming a new virtual homebuyer invested in people who, in return, will invest in you and your purpose.

Justin Cappon Pro is a brand strategist who stays on top of the latest and greatest virtual properties. He welcomes you to visit his crib, "World Changers Mastermind," on Clubhouse anytime you are strolling through the neighborhood. His handle is @justincapponpro, and he would love to meet and greet you.

Social media is a very, very powerful tool. It also gives power to tools.

CHRIS YOUNG

Hack Two

AVOID ADVISORS

Although I had a great realtor who is very good at what he does, I also had a few advisors who almost convinced me not to seal the deal on this virtual property. Always do your go own research. When taking advice, avoid getting too much information from those who have no experience in the industry.

My realtor is very knowledgeable about virtual spaces. He explores new apps all the time, setting up accounts, exploring the benefits, and learning the downsides of spaces. He crunches the numbers and tracks the trends before conveying any knowledge to keep his clients from jumping to the latest or greatest whatever. His diligence made me comfortable to at least consider his invitation to tour the Clubhouse community. I trusted his research after years of accurate and helpful advice.

Some advisors don't want you to move simply because it isn't what is best for them, though it may be the best move for you. Change is difficult for everyone, but change challenges us to seize what is possible. Challenges can change your perspective to see opportunities you might have otherwise overlooked. Everyone won't agree with your decision to change, and that's okay. However, make your decision based on the knowledgeable, research-based knowledge of a trusted advisor.

At the beginning of 2020, my husband and I attended a black-tie charity event benefiting children with type two diabetes. We were very optimistic about the year ahead and all of our professional and personal development plans. My husband has worked as a First Responder for over 18 years and started a business teaching safety to civilians, including First Aid and CPR certifications. He really enjoys teaching practical skills and relating them to real-life experiences, as do I. Neither of us (and most of the world) had no idea we would have to adjust all of our plans and expectations just a couple of months into the new year. Towards the end of April, my husband reminded me that there would be two kinds of people coming out of 2020: those trembling or those thriving. What we chose to do with the challenges of the Coronavirus pandemic would make the difference. Regardless of all the loss and terrifying uncertainty, we decided not to live in fear. We immediately shifted gears and decided to invest in finding those with knowledge on overcoming challenges and converting those into successful steps toward success. Many advised us to should brace ourselves for the worse, slow our goals, and hunker down. Then, there were those with the knowledge we needed to challenge us to do the complete opposite. Fear definitely was a factor. However, the fear of failure or getting locked in loss was not greater than our decision to thrive as we survived. Unfortunately, I did lose a family member to Covid19. Several of our friends suffered major losses while enduring the challenges of 2020. Our determination was not in disrespect to those who suffered. Despite the tragedies around us, we chose to press forward and honor those loved ones and friends. With this mindset, my husband and I took a chance at positioning ourselves to do something for those we care about in the future. We remained open to possibilities. The challenges we faced changed our perspective.

Witnessing what the world was going through and not being in a position to do much about it was heartbreaking. It built up our resolve to push ourselves harder than ever before. I didn't know that the result of that push would lead me to Justin Cappon and, eventually, Clubhouse. At that time, the virtual spaces I found myself in didn't allow me many opportunities to let my voice be heard. It was a struggle to get past my immediate circle of influence. I needed an expert to help me, and it wasn't going to come from those in the same situation I was already experiencing. That is why this hack is so vital. We've all heard the saying, "Be careful who you listen to," and there is merit in that advice. You can't always go with what you know; our current knowledge will only keep us in cycles of our recent results.

Yes. Most of the time, it will cost us something to learn or gain what we don't know to get different results. In 2020, I hand drew and published an adult coloring book to help those needing a break from all of the stress of their day. I worked with a young aspiring author on my first children's book; she illustrated all the pictures. All the proceeds from the sales of that children's book go toward mission projects in Africa that I wasn't able to support at a high level. Finally, I was also able to complete and publish the final and third volume of my young adult, Christian, fictional series. My husband found a way to continue to support those in need of CPR and First Aid certifications. I share our accomplishments to demonstrate the difference advice can make from someone with wisdom and knowledge in their field. The benefits of wise advice is priceless for you and the perspective you have during challenging times. Instead of slowing down, my spouse and I did the opposite.

More than a hack, I simply suggest you learn from my hindsight on 2020. Don't be discouraged by challenges, fear, or the advice of those inexperienced in what you want to accomplish. Get in the company of those who know what you need to know to move forward. Refuse to listen to those who are simply speaking from fear or a lack of understanding. If you want all that life has to offer, you must look at the world around you and understand that it is changing at a rapid pace. If you get these hacks under your belt, you will navigate the changes much easier, and your new perspective will be priceless.

What is interesting is the power and the impact of social media. So we must try to use social media in a good way.

MALALA YOUSAFZAI

Hack Three

KNOW THE VALUE

After deciding how much home you need and can afford, you must explore your potential property's value before making your final decision. Yes, this applies to virtual property as well. You don't want to get invested in something that isn't worth your hard-earned resources and time. To explore the property's value, you must do a walk-through and get an appraisal of the property. Have you ever considered the property value of your virtual space? Did you do a walk-through of the spaces you currently occupy, or did you move in assuming all risk? Consider the impact you will have on the space and the impact that space will have on your lifestyle and development goals.

If you were to move from the current space you occupy, would there be an impactful loss to that community or yourself? Would the value you offered to that community be missed? Would you miss the value the community offers you? These are important questions to ask yourself while considering if the trouble of moving is worth it. Picture yourself in this new space and consider the impact you may have on this new community. Do you see value in that?

Your realtor can walk you through the property and have an appraiser give you the rundown on the current value of the space. In the walk-through, your realtor and the appraiser should be happy to answer any questions and explain anything you are unsure about regarding the community.

When I conducted my walk-through of Clubhouse, it was in beta testing, which means it was still under construction. The value of this property had to be determined by the investors backing the construction and the community wanting to get in on the ground floor of this influential space that could potentially become the best virtual neighborhood in the world. I was introduced to an opportunity to contribute and impact others in a manner that I had not experienced before.

During the walk-through, I pictured having cookouts with the Smiths and potentially meeting a celebrity or two on my morning jog. I would be a regular attendee and participant of the neighborhood association to do my part in keeping this new community safe and thriving. The value of being a part of this community was adding up with every step I took throughout the various rooms of the Clubhouse. Did I mention the sun was shining brightly the day I did my walk-through? I would have to come back on a cloudy day and walk through it again to be sure I valued these perks the same under different circumstances. Walk through the property more than once under different circumstances. Only you know what feels right to you.

Clubhouse broke virtual ground in 2020 during one of the most trying years of my existence. It consisted of a community of survivors seeking support and refuge from a world of social distancing. I stood in the middle of an empty house without walls and imagined a place I could call home. The plans for this space were being decided in the moments I found value. You must walk through the spaces of your life and measure the value against what is being offered. Will this space support the growth of where you are headed? For me, Clubhouse was proving valuable for my current and future development goals. These goals affect me personally and professionally, and I was finding the value I had been seeking.

I could go on and on about the many values I found in Clubhouse; however, I will spare you and only share a couple that I discovered. What is valuable to me may not be the same for you. I am sharing my experience because, like my realtor, it has been too good of an experience to keep to myself.

The first thing I noticed when I conducted my walk-through was the freedom Clubhouse allowed me. Many other social media spaces don't allow you to keep the app's features running while functioning outside of the app itself. With Clubhouse, I can go in the house, sit in my favorite room, and listen while taking a walk to my e-mail box or checking the weather channel online. I can visit other platforms and still be tuned in to what is going on inside Clubhouse. This exceptional feature allows me to consume knowledge while also offering me the ability to interact, ask questions, or remain silent while going about my day.

The ability to access other things that support my day without the app shutting down was a definite value. The second thing I noticed was how hard the contractors worked to get the minor details right for my individual needs. If I need more space in my office instead of the kitchen (I'm a horrible cook), they were able to accommodate and even suggest things I didn't consider. Of course, I don't mean a literal kitchen or office, but similarly, the virtual space you are moving into must offer a better experience or cater to what you have been used to at a minimum. Believe me when I tell you that this space is not a cookie-cutter space with the typical features and benefits. Therefore, its value is unlike any other virtual space I have ever been apart. The architects of Clubhouse work tirelessly to provide upgrades and amenities tailored for the potential buyer, not the developers. This Clubhouse space is created for the creative, and I felt at home before I met the first neighbor.

The best value I experienced during the walk-through was the diversity of the neighborhood. First off, I was looking for a little cartoon pic or fancy font for the Clubhouse icon sign. It was like crossing the state line of a state you have never been to before and finding out they are known for their cheese. It causes you to slow down and ponder that for a moment. I was pleasantly surprised that there was a person of color's face as the app's icon. It was very cool that the app's builders utilize the faces of real people and real users/residents to represent Clubhouse. I had never seen anything like it on any other social media app before.

Most of the neighborhoods I found myself within other platforms did not look like me, and the only time I saw others like me were the times I invited friends or family over for a visit. Those other spaces also made it difficult to intermingle with others who didn't look like me if I didn't already have them as "friends" or "family" in that space. However, Clubhouse is very different. It is a community of inclusion, and it is visible from the moment you locate the icon and download the app. There has been equal opportunity to join this community from the first day the builders broke virtual ground. I hope this is something you consider a value as you begin to explore the benefits of moving into any virtual neighborhood. It truly has been a value to me to have access to so many beautiful, culturally different communities without having to do more than jog down my virtual street.

The builders of this app keep the buyer's needs in mind instead of the investors. Around the Clubhouse neighborhood, you'll often hear people say that the purpose of this space is to have and foster great conversations. In a world starving for connection and effective communication, there is no better place to take up residence than a place that makes genuine connection their primary value proposition.

Have you felt the need to move yet? Are you beginning to see the value in being in a community that supports you and works hard not to suppress your inner creativity? I encourage you to take a walk-through and build a value list of your own. It will come in handy when you begin to determine if any virtual space is a good fit for you.

Hack Four
HIDDEN COSTS

At this point, you may be thinking this all sounds too good to be true. Somehow I may have convinced you that Clubhouse is a virtual Utopia that everyone should join. Let me caution you regarding the hidden costs and fees. With any investment, even those that pay off huge dividends, you must consider typical fees and hidden costs. Virtual real estate is no exception to this rule. Know the hidden costs and fees before you go all-in on any social media app. You may like the neighborhood, and the home is everything you could want and more; however, the sticker price isn't the only cost involved.

My realtor warned me about the hidden costs and fees, which he equated to my time and contributions to the app. Clubhouse is not responsible for these hidden fees, so they are off the hook. The hidden costs and fees are determined by your virtual lifestyle, which sets your cost of living and interaction within any app, not just Clubhouse. Only you can determine what those hidden costs and fees will ultimately look like. The impact it has on your current or future lifestyle is entirely up to you. Make no mistake. They are there, and you are paying them whether you realize it or not.

Before Clubhouse, I would categorize my virtual living arrangements to resemble that of a renter. I never took true ownership of the virtual space I resided in. I joined social media platform after platform, looking for the right fit. Time after time, I discovered that as I moved into these spaces, the virtual neighborhood's culture had already been established. I didn't have much say on how I wanted my specific home to look or feel within the community. There were limited options in how my entry into the space could potentially contribute impact.

Before joining Clubhouse, I was used to paying my rent, following the rules, and accepting the limitations of my residence. Therefore, I didn't spend much time or resources in those kinds of spaces. The hidden cost involved in being a part of the platforms I'm currently or formerly a part of varied because of how I utilized them. The main thing to remember is that none of the virtual spaces force you to do anything but your own life choices will set the cost of the hidden fees. Since moving into the Clubhouse app, I have quickly found out the real costs of finding a virtual home you can take ownership of. Everyone knows there are more costs involved in home "ownership" as opposed to other temporary arrangements. I fully understand that you and I will never actually "own" Clubhouse or the space we occupy within it. However, in the virtual world, it is about as close as you will get at this current time, and I think they are setting the pace for what is to come.

Although it has its costs, homeownership comes with so much value that makes the cost worth the reward. Some may argue that the cost is too great and again, that is something you will have to decide. Many have become the gurus of other platforms. They found their niche within it, and that has made all the difference for them. I am not knocking that experience whatsoever. I am specifically addressing the many others who have not found their forever virtual home. They have not found a social media space that provides the freedom to choose how they want to spend their time online. That was the case for me personally. As a creative, I had yet to find a space that appreciated my needs and took time and effort to make those needs the center of the app's functions. Clubhouse was created with the creative in mind first. Wow! Just saying that gives me visuals of a blueprint with all my customized requests to make my living space accommodating to my wants and needs. For this kind of customization, I am willing to pay the additional fees and costs.

When long-distance came along for the average

telephone user, it was insane to think you could speak to a loved one or friend across the world. Living as a dependent in a military household made this function invaluable, as families could connect no matter where they were stationed on the globe. I think of Clubhouse the same way. It provides a mutual meeting place that we all find valuable and allows for a closer connection. The best part is that I don't have to have known you ahead of time or go digging for your phone number. Similar to long-distance calls, having additional costs and fees depending on where you were calling, the function of connecting with people all over the world on Clubhouse will have its fees also. You may sacrifice some sleep as you choose to connect with someone in a different time zone on the other side of the world. For me, it is a cost I'm happily paying.

I had to add other apps to my phone to keep connected with those in places where regular texting or messaging methods costed more. I had to decide how many apps and how much extra effort to stay connected I was willing to make. These extra costs, including the time we spend on Clubhouse, may seem a bit much to some, but I respectfully remind those who say this that you are probably coming from a virtual space where you have already made these same kinds of investments, and you are happy there.

It is easy to drive through a neighborhood you

don't live in and decide what is too much or excessive for those who found a forever home and lifestyle they love. I challenge everyone to imagine wanting to be in that kind of neighborhood and space, but you hadn't found it yet. You would spend the time and gladly pay the cost and fees. I admit that Clubhouse won't be the right fit for everyone, but I implore you to find your fit in this virtual world.

Last, I want to address the cost of time. If you were having a custom home built, how much time would you spend driving by and overseeing the progress? You would probably spend every free minute going by to see how far the builders have come with the living space and then back the next day to see if they have started on the kitchen and so forth. You wouldn't stop checking in and going over and through the plans until move-in day. Then, the real-time spent in and on the home begins. It is the same thing for Clubhouse and those working to build something we can be proud of and possibly call home. If you already have your home and virtual real estate is old news to you, bring a housewarming gift to your new friends experiencing this for the first time. Give their neighborhood the respect it so richly deserves for providing a place the creative can call home.

The best advice for anyone trying to grow on social media is to be real and to be consistent. People will connect more with you if you stay true to yourself and who you are.

LELE PONS

Hack Five

NAIL THE LOGISTICS

You are now experienced in dealing with a virtual realtor and how to evaluate good virtual real estate, using the Clubhouse APP as an example. It has certainly been a great experience for me, and I wanted to share this experience so that others will have more fun and less stress finding a virtual home. Once you've found your perfect virtual space, it's time to get down to the logistics of moving into your new place. Whatever you do, don't put off packing the essentials. The advice in this chapter is catered towards moving into Clubhouse, however, you can use this wisdom when moving into any virtual space you've decided on.

We know the space is right for you, and that is why you are moving in, but remember that this space isn't like your old one. You may need to get rid of a few things from the old place. Not everything will fit into this new home with the new layout. Some things won't fit because it just doesn't go with the new you and your décor theme. That's right. You get to have a fresh start here and decorate this home the way you want. It doesn't have to be virtual living as usual! Start with the essentials first and work logistically from there. For example, your bio from another platform may not fit on Clubhouse. You have to determine your why for this space almost immediately to begin moving things into place with intentionality. What was it about this app that spoke to you and made you feel it could be your forever home? For me, it was the opportunity to get into the space on the ground floor, have influence and impact in the community, and experience the cultural connection and phenomenon of organic relationship building. Whatever your reasoning, use it to move into your space with intentionality.

To avoid the stress that can come with figuring out the logistics, try this little hack. Check out the bio of the person who invited you to the app or your virtual realtor. Look at what they have, use it as a guide, but adapt it to represent your purpose for being on the app. For example, I am an author, speaker, and artist. On other platforms, I focus primarily on speaking. While on yet another platform, I may focus primarily on the written or artistic side of what I do. Therefore, those same bios may not work for me on Clubhouse. I had to decide early on what kind of home I wanted. It is as detailed as choosing a high rise in the city or a log cabin in the woods. You have to know the purpose of the home you just invested in and how it supports your lifestyle. This will lessen the stress as you begin to work on the logistics.

For me, Clubhouse is a large family-style home with many rooms to include a craft room and a wrap around porch where I do most of my writing and drink coffee with friends. The furniture is not covered in plastic, and it is minimal to allow room for entertaining friends and space to create the things I love. To do this effectively, I make sure that my layout is warm and welcoming.

Everyone is welcome to stop by and sit with me awhile on the porch swing. You may even be invited to come inside and find a comfy spot to snuggle up near the fireplace. Don't worry about asking for anything because, in my home, the saying is "Mi casa es su casa" or my home is your home. Help yourself to whatever is in the fridge and pantry. We play board games at this house and share things that can help us get further in our careers and life. We may open that Bible on the coffee table if things get too deep, but everyone is welcome in this house. Come as you are. The logistics of my virtual home support my efforts to build meaningful relationships while providing a place for me to grow personally and professionally.

Logistics involve more than just the layout of what will go where. Logistics are a coordinated effort to organize the details of your new living space to include the many people involved and supplies necessary to make this space as efficient as possible. Apart from your realtor, you may want to consider who to bring along on move-in day. This should be a trusted friend or close relative that knows you well enough to know what supplies you may be missing to get comfortable. On a virtual app, supplies can look like knowing the technical side of the app and how things work. The logistics involving people may include who to follow as well as paying attention to who is following you. Don't fall into the trap of just following everyone that follows you or being bullied into following others at the suggestion or threat of someone else. Unfortunately, no virtual space is free from the neighborhood bully. No worries because the bullies on Clubhouse are outnumbered. Unlike other apps, the neighborhood association has guidelines in place that are strictly enforced. Clubhouse is constantly being upgraded to make the residents not only feel comfortable but safe.

Having a friend or family member work out the details will help you make decisions that lessen or possibly avoid stress. If something doesn't fit the picture you created for yourself and your new virtual home, it doesn't belong in your virtual home on this app. The key is to allow the logistics process to free you, not overwhelm you.

I have a LOT of experience with moving into new homes in the real world sense. My mother married into the military, and we moved quite a bit. My mother came from a time when you didn't put off tomorrow what could be done today. This was a good thing because no matter how many times we moved, the new place felt like home in what seemed like an overnight process. My mom would have the whole family up until the early hours of the morning unpacking, and before lunchtime the next day, the only sign that we had just moved in was the flattened boxes at the end of the driveway. I share this story because that was a different time and definitely under different circumstances.

If you feel like you have to have everything in its place and all the logistics figured out on move-in day, you are certainly mistaken. The Clubhouse app is constantly doing upgrades. There are gurus for the app already in the making, figuring out all the tips and tricks to make your experience easier. For now, all you have to do is move in and unpack as your family and friends help you work out the logistics.

Don't feel like you have to have everything in the right place or the right room right away. You shouldn't put off unpacking for a long time when you get into this space, but don't get overwhelmed by what to do first and by when.

Do a deep clean before you start to unpack. Creating and unpacking your bio is similar to forwarding your mail from your old address. You don't want to skip this step or wait so that your family and friends know where to find you and stay in contact with you. It also allows your new neighbors and friends to get to know you better and learn the best way to connect with you. Keep in mind this is a new address and a new opportunity to start fresh. Make this home all that you want it to be for where you are now with your life goals. Old rules from the old neighborhood and culture do not apply here.

Hack Six

CHANGE THE LOCKS

This hack is all about protecting yourself and your new virtual home. Just like you would do when you purchase a new physical home, you want to change the locks immediately. On a virtual platform, protection means putting measures in place so that you lessen the chance of exposing yourself to unnecessary dangers. You should get in the habit of checking the profile/bios of those you wish to follow. Visit your list of people you follow periodically and reevaluate if you need to remove some individuals that are not contributing to the virtual lifestyle you wish to have. Don't worry about removing people from this list. Clubhouse is a safe place, and retaliation for removing someone is not frowned upon. You are in control of your virtual living space. Do what you need to do to make it as comfortable and safe for you to experience the best the app offers.

Locate the fuse box so that you know where to go in case of an extreme emergency and have to shut things down quickly. On Clubhouse, the fuse box is equivalent to the block feature on the app. If someone comes into the room, there is a feature where you can tap on the individual and block them. This sends a notification to Clubhouse to keep an eye on this individual or investigate them for disruption on the app. This should be taken very seriously, like going to the fuse box of a physical home. You don't shut this off because you don't like the utility bill. Use the block feature with caution but most definitely if the situation calls for you to do so. Clubhouse takes bullying seriously enough that if you block someone, the person who invited/nominated that individual onto the app can also be investigated as well as possibly kicked off the app.

At the end of the day, your safety and comfort are what matters. However, don't confuse comfort with an actual emergency. I am married to a first responder of over eighteen years, and the number of false calls and false emergencies he has received would surprise you. Don't be that person to use a "safety" feature as a tool for revenge. I hope that doesn't require more explanation.

I won't go into various tips, tricks, and app features because the app is in beta testing as I write this book. Things will probably change over the coming years, and I'm looking forward to witnessing Clubhouse's innovative improvement journey. What doesn't change is the need to know how to protect yourself online. Whether you are on Clubhouse or any other virtual app, locate the options you need to protect yourself.

Unfortunately, although rare, bots and trolls have found their way onto the Clubhouse app. I've only seen two instances since my time on the app, but it is easy to mute individuals who come on to your platform with ill intentions. The other option is to return that person to the audience (you will learn more about app features when you take advantage of hack #1). If someone isn't participating in the conversation that lends toward the topic at hand, you can mute them or return them to the audience as well. Remember, this is for extreme cases when one individual is not respecting or keeping the comfort of all houseguests in mind. Use these emergency shut-off systems to protect yourself and others in your Clubhouse room as necessary.

Another way to protect yourself in any virtual space is to have the same guardrails you would have in a physical location. Those protections include having door greeters to welcome those coming into your virtual space. They let the visitor know that they have been seen, which can be a good thing for both the host and the guest. The guest feels important because someone is paying attention. You feel safer because you have greeters in place who are "paying attention." Just be aware of who is in your space.

Clubhouse is great for connecting and meeting new people, and you want to experience this part of the app as best as possible. As you get into the app, visit rooms, and possibly start a room of your own and invite guests, having a couple of door greeters is an excellent protective measure. These can be people you assign to recognize visitors and returning friends and family to the room by name. How you structure this is up to you, but the key is staying aware of your environment.

Another protective measure is having a "bodyguard" who sits in the room and acts as security for the space. This doesn't require a lot of fanfare or attention. It only needs to be someone willing to step in if a visitor/speaker is getting the room off track or taking the conversation down a road that makes you or your guests feel uncomfortable. The "security guard" has permission to interrupt this person as respectfully and tactfully as possible. Security would then redirect the conversation back to the original topic of the room. How security redirects those who get off track or start to stray from the core theme of your space is totally up to you; however, it is an invaluable protective tool you should have in place, especially if you run a room of your own. A room is simply the space you open up on Clubhouse to host great conversations and facilitate talks around various topics. Again, I won't go through all the specifics of the Clubhouse app, you'll learn more as you get involved in your new virtual community if you've made Clubhouse your home.

The great thing about these safety measures is that it takes the pressure off of you to be all things while contributing value to others or receiving information others want to share. Having these measures in place will allow you to operate in a safe place that fosters great conversation and generates organic relationships that will change your life and the lives of those you meet. During my time on Clubhouse, I have met many new friends and developed genuine relationships. I can't say enough about Clubhouse and the connections that have blessed me in such a short amount of time.

My experience on other platforms taught me that I should not expect genuine connection online. On other platforms, I would be plagued with ads, stalked by trolls, and bombarded by spam in my messages. I know that Clubhouse will have its issues, but the app does its best to avoid those that I have just mentioned. As a result, I'm free to focus on building relationships that matter. Connecting is easy on Clubhouse, so utilize Hack #6 and connect safely.

There is no debate that social media is a great tool for networking with others in our industry. It can lead to friendships, support, and serendipitous connections with reviewers, agents, reporters, or editors.

M. J. ROSE

Hack Seven

MEET YOUR NEIGHBORS

Hack #7 is my favorite! This hack is all about connecting with amazing people in your new home. Hack #7 requires you to be open to meeting new people and building authentic relationships. If this scares you or makes you uncomfortable, you will inadvertently block the very experiences that will make Clubhouse a rich and thriving neighborhood to live in virtually. However, if you follow all the hacks in this book, you can feel confident moving into any virtual space with confidence.

Now, what kind of experience would this be if I just told you, "Go meet new people" and didn't provide you with a few suggestions? In this hack, I will provide you with the addresses of some of my favorite neighbors on Clubhouse. If I could afford to share the full story behind the individuals I'm about to introduce you to, that would be another book in and of itself. I won't ask you to read all of that, but I will invite you to find out more about my neighbors on your own. Interview them, bring them on your podcasts, join them in a room or ask them to help you with your room. I hope that you look them up and connect with those who resonate with you on some level. Feel free to borrow a cup of sugar or get directions to the nearest grocery store. They are happy to help, and if a couple has moved since you came onto Clubhouse, no worries, there are many wonderful people on this app waiting to welcome you into the neighborhood.

I've already introduced you to my realtor, Justin Cappon Pro, who runs the World Changers Mastermind group. This group welcomes anyone interested in working with a peak performance coach with lots of brand strategies in a mastermind setting. Justin is the reason I was introduced to Clubhouse and is always welcoming newbies to the app. Look him up, and I'm sure he would love to have a chat with you. Those in his Mastermind group will be excited to welcome you as well.

Charlotte (@WebiStebi), a Holistic Life Strategist, Celia (@celiakibler), an expert Parenting Coach on a mission to stop 1 Million parents from yelling at their kids, along with Keith (@dads4life) are all members of the World Changers Mastermind group. Larissa (@Gr8tfulChick), also a member of this group, has a powerful story and considers herself the C.I.O. – Chief Inspiration Officer of the Gr8tful Tribe. The following are also part of Justin's Mastermind group:

- Maggie (@maggiegraham), a Homeschool Mom of 11 kids
- Shirley (@shirleyspeaks), an author, publisher, and writing coach
- Kate (@katgartung)

Once you move in, join this dynamic group for some front porch conversations about business and the life we live behind the scenes. We would love to get to know you better.

Most of the following neighbors I'm about to mention, I met as I began to live my virtual life on the Clubhouse app. With my husband and a close friend as an exception, I did not know any of these individuals before December of 2020, when I originally joined the app. They are a testament to the connection and genuine relationships you can build if you are open to the opportunity. I truly am grateful to each of the following people for making me feel right at home.

As I mentioned, my husband is one exception to the statement above, so I will start with him. Chad (@safetyexchange) is my husband and a First Responder in the physical world, so consider him the neighborhood safety guy on this app for all your First Aid, CPR, and safety concerns or questions.

My close friend, who I also mentioned as an exception, is one person you will want to get to know and follow, especially if you need a good prayer warrior on speed dial. Ivery (@Iveryyarbrough) is a competent speaker, a graduate of Oral Roberts University, and an amazing make-up artist. She can whip any face into shape, including mine. The beauty she reveals comes from her ability to pull from the inside what is truly beautiful about a person and lay it on display. I hope you look her up or check out her podcasts.

When I first moved into Clubhouse and didn't know very many people, I happened to go by the gas station to fuel up and met a gentleman named Alpha$ixx, who has been an encouragement to me on the app ever since. He introduced me to Danna (@ dannaklicko), whose name rhymes with Nanna, and she, along with Alpha$ixx (@juanjordan), volunteered to show me around until I got comfortable and began to meet more friends in the neighborhood. When they are not adding value in other Clubhouse areas, they both visit from time to time and show their support for all that I do. Alpha$ixx is a self-actualizing human being who challenges us to think thoughtfully and with intention. If you want that kind of challenge, you should look him up.

Next up is the local owner of the A.o.A, Nathan (@nathankettler), also known as the Alchemist of Awesome. Many hang out with him in the A.o.A. center. I can't say enough about Nathan and this band of awesome misfits and major rock stars.

Nathan is often seen with Emily (@nativedeathdiva), who helps keep him organized and offers her expertise. This dynamic duo and all those who hang out with them in the A.o.A will help encourage you in all your efforts to make an impact on the world. Nathan advocates for his crew to squad up, and they definitely take on the challenge. I met Nathan through another awesome guy on the app who is all about purpose and connecting like-minded people. His name is Joey (@joeychandler70), and he will not only connect you to awesome people like Nathan, but he helps you move forward in your purpose in such an intentional way. Joey has a podcast in which he graciously invited me to be a guest. Joey will get you motivated and moving forward in your purpose goals.

There are many Clubhouse members connected to the A.o.A. you must meet because they will not only lend you a cup of sugar, but they will take you to the store and buy you a whole bag of your own. Yes, they are that sweet! The following are all connected to the A.o.A. community:

- Julie (@traxleris1)
- Brandyn (@mrcottingham)
- Kate (@kategarneslive)
- Dr. Davia (@askthedrd), who is an author, has shown up for me and given me a virtual check-up on my goals alongside the advice of Nathan, who

truly invests in all those who support the A.o.A.

I can't forget to mention the one and only Rock Florida (@rockflorida), the most sincere friend you will find on Clubhouse while demonstrating loyalty to his family. I could say more about each of these individuals, but you will have to look them up and find out for yourself.

You will be fortunate to spend time with everyone and especially Nathan, the Alchemist of Awesome, who supports his Growth Squad in authentic awesomeness in their business goals. This group is like visiting the gym. You will get a workout when you go and be better for it, and when you miss a day, the results begin to show. Don't skip the A.o.A room!

I met another individual at the virtual library who gave me a membership card the first day we met. He invited me to participate in a dynamic book club and other educational rooms that support driven, heart-centered people looking to make a difference. His name is Dave (@daveburlinand). Books are not his only jam; he is a DJ, a speaker, and a former member of the military who deserves our gratitude. Read Dave's bio, and you will find out just how special it is that he has not allowed his accomplishments to separate him from genuinely connecting with people. If you are ever fortunate to cross paths with Dave, it will be one encounter to remember.

If you are an early riser and looking for inspiration

on how to live life with affirmations, you need to Declare the Morning with Kim (@visioncoachkim). Coach Kim has a room full of people learning to declare the things they want for their lives every morning according to scripture and science.

There is a dance studio not far from my residence, @itsanewday365 by Kenyetta. The energy from Kenyetta is evident the moment you bust a move through her virtual doors. Next door to her is a dear friend everyone calls Coach Paul (@setfreecoachwho) is about all things freedom. Coach Paul is one of the most sincere people you will meet who genuinely has your best interest in mind when he interacts with you. He wants to see you succeed!

Learning your way around a new neighborhood can be a challenge but not if you open up and ask questions. I asked questions and asked for directions when I was getting lost. I found that people were so willing to be helpful and assist me in making the right connections that supported the life I wanted to have on this app. Among those I've met are the fabulous ladies at the Idea Machine. I met Jenn in the A.o.A. group, but she invited me to visit her friends at the Idea Machine. Jenn (@byjennsteph) helps women feel beautiful inside and out. She is often in the company of Mario of the Idea Machine @botanically_blended, sharing her knowledge of business and marketing tips over a cup of tea. There is also the awesome

Imani (@imimaniallen), making a significant impact in the businesses of women who need to get beyond their limiting beliefs and systems.

I must also mention the queen of influence herself, Neesha (@neeshasnetwork), which is appropriate as she is the networking master. Neesha does much more than I can even begin to explain in a few short sentences, but she is genuine in her efforts to bring people together. I am still getting to know Neesha, but she has treated me like a sister from another mister from the day we met on Clubhouse. She has used her networking skills to connect me to people and resources I needed to know about. You may want to check out Neesha's Network!

As I began to take my daily jogs, swing by the library to pick up a book from Dave, circle back to the A.o.A for some daily encouragement with Nathan after spending some time with Justin and the World Changers Mastermind group, I ran into a financial advisor named Troy (@troyholt). Troy introduced me to Angela (@naileditempower). Both Troy and Angela have welcomed me to the neighborhood like I was family. Troy is passionate about helping people get out of debt and manage their finances well, while Angela is the very definition of driven. Angela will give you the business breakdown and do your nails at the same time. Both Troy and Angela frequently invite me to assist in their rooms and helped me to get comfortable moderating on the app. You definitely want to go by Troy's office if you are serious about financial literacy and relationships. Angela isn't far from his office, so stop by and spend some time with her as well.

As you can see, my neighborhood is not only diverse but is very welcoming in all aspects. Each person I have met on Clubhouse is not listed in this book. This chapter doesn't begin to exhaust the individuals that have contributed to my personal and professional growth since being on the Clubhouse app. I couldn't possibly list everyone I associate with, so I had to pick the few that I visit on a daily or weekly basis that quickly approved my mentioning them in this book. I hope you take the time to visit them when you do your walk-through of the app or eventually find yourself moving into this virtual space.

There are just a couple more people I want you to meet. My friend, Professor Nez (@professornez), inspires you to reach your greatest purpose. He is as authentic as they come and loves sharing his other platform passions. Learn how to make the most of Clubhouse alongside his favs and then join his room to hear what he is discussing.

Last but certainly not least is the publishing queen, Jessica, who has her own book publishing company called Keen Vision Publishing. You can connect with Jessica (@meetjdwilliams) and start your journey to publishing a book of your own. She and her crew are responsible for helping me share my Clubhouse move-in experience with you. She is another exception, in that I knew her before moving into Clubhouse, but I didn't know she was in the neighborhood when I moved. Have you had that happen before? You move into a place and find out a friend or acquaintance lives nearby, and it is a pleasant bonus to living in the area. Before you know it, your community of neighbors will be as vast and rewarding as mine. I don't mind sharing my neighbors, but please be respectful and kind to the people that have been respectful and kind to me as I've navigated my way down the streets of Clubhouse.

If I didn't mention someone's name and handle, it is not because they are not awesome or have not been impactful regarding my experience. My sincere thanks to all those who have encouraged and inspired me to share my move-in experience.

If you decide that Clubhouse is the virtual home for you, I would love to get to know you! My home address is @puzzled2purpose, and I will happily welcome you to the neighborhood. I can introduce you to the gurus of the app with all the tips and tricks, but not before you come over for dinner.

NEIGHBORHOOD DIRECTORY

Alpha$ixx @juanjordan *Knowledge Broker*
Angela @naileditempower *Nailed It Empowerment*
Brandyn @mrcottingham *Investment Design*
Chad @safetyexchange *CPR-First Aid Certifications*
Charlotte @WebiStebi *Holistic Life Strategist*
Celia @celiakibler *Expert Parenting Coach*
Coach Paul @setfreecoach *Set Free Life Coach*
Dave @daveburlin *Las Vegas-Speaker-Podcast Host*
Danna @dannaklicko *Essential Oils – Natural Living*
Dr. Davia @askthedrd *Author- Your Best CH Wing Person*
Emily @nativedeathdiva *Native American, Death & Grief Education*
Ivery @iveryyarbrough *Biblical Educator,*

Influencer, Evangelist

Imani @imimaniallen *Creating Significant Impact in Business*

Justin @justincapponpro *Startup Business Mentor*

Jenn @byjennsteph *Engineer Beauty Guru*

Joey @joeychandler70 *Purpose Coach*

Julie @traxleris1 *Small Business Expert*

Jessica @meetjdwilliams *Keen Vision Publishing*

Kenyetta @itsanewday365 *Podcast: It's A New Day*

Keith @dads4life *Dad Coach, Author, Blogger*

Kate @katgartung *World Changer Mastermind Member*

Kimberly @visioncoachkim *Mindset and Business Coach*

Kate @kategarneslive *Turning Humans into Gladiators*

Larissa @Gr8tfulChick *Chief Inspirational Officer*

Maggie @MaggieGraham *Homeschooling Mom of 11*

Mairo @botanically_blended *Botanical Tea Empress*

Nathan @nathankettler *A.o.A –Alchemist of Awesome*

Neesha @neeshasnetwork *Life Coach –*

Networking Enthusiast
Professor Nez @professornez *YouTube Pro, Branding Agency*
Rock @rockflorida *Real Estate with No Realtor*
Shirley @ShirleySpeaks *Author, Publisher, Writing Coach*
Troy Holt @troyholt *Financial Strategist - Encouragement Officer*
Tasha @puzzled2purpose *Purpose Advocate, Author, Speaker*

There are not egomaniacs listed in this directory. We are all here to help.

Your piece matters.

TASHA HART

ABOUT THE AUTHOR

Tasha Hart was born in Terre Haute, Indiana. She has lived in Tennessee for over seventeen years. Tasha has an associate's degree in Biblical studies and continues to find ways to share that knowledge in all that she does.

She has consulted hundreds of organizations in achieving their non-profit 501(c)(3) status and using that status to impact the world. Tasha enjoys connecting with others on all levels in order to share her message, "Your Piece Matters." There are many pictures in the universe, and Tasha believes every person has the responsibility to find their fit among them. She has earned the title of a Distinguished Speaker and shares the message of purpose and her Puzzle Philosophy in a variety of settings from ministry groups to corporations. Tasha has won various speaking awards including first place in the 2020 District 43 Toastmasters Speech Evaluator contest.

Currently Tasha is the author of several books, Puzzled by Purpose and The Secrets of the Twelve Trilogy. All the proceeds of her children's book, Eddie and the Rain Barrel, support mission projects in Africa. Behind every book Tasha has written, there is a genuine desire to connect with others and impact their lives for the better. When she isn't writing, coaching, speaking, or assisting others with their projects, Tasha paints wooden puzzle pieces to express inspiring thoughts and ideas. Many have purchased custom pieces to honor those in their lives because "Their Piece Matters."

Tasha currently resides in Jackson, TN with her husband and First Responder, Captain Chad Hart of the Jackson Fire Department. They have three children between them with a large dog on the way.

STAY CONNECTED

Thank you for reading, *Keys to the Clubhouse.* Tasha looks forward to connecting with you. Here are a few ways you can connect with the author and stay updated on new releases, speaking engagements, products, and more.

YouTube	Find Your Purpose w/ Tasha Hart
Podcast/Spotify	Find Your Life Purpose w/Tasha Hart
Facebook	@puzzled2purpose
Twitter	@puzzled2purpose
Instagram	@puzzled2purpose
Clubhouse	@puzzled2purpose
HAPPS	@puzzled2purpose
Website	www.puzzled2purpose.com
Email	tashahart@puzzled2purpose.com

Made in the USA
Monee, IL
04 November 2026

67380165R10036